THE HUMAN BODY
Let's Investigate

by Ruth Owen and Victoria Dobney

Consultant:

Nicky Waller

Published in 2019 by Ruby Tuesday Books Ltd.

Copyright © 2019 Ruby Tuesday Books Ltd.

Editor: Mark J. Sachner
Designers: Emma Randall and Tammy West
Production: John Lingham

Photo credits:

Alamy: 5 (centre), 28; Science Photo Library: 4 (top), 9 (bottom); Shutterstock: Cover, 1, 2—3, 4 (bottom), 5 (top), 5 (bottom), 6—7, 8, 10—11, 12—13, 14—15, 16—17, 18—19, 20—21, 22—23, 24—25, 26—27, 29; Superstock: 9 (top).

ISBN 978-1-78856-034-4

Printed in Poland by L&C Printing Group

www.rubytuesdaybooks.com

Contents

Our Amazing Bodies..4

Structure and Strength..6

The Digestive System 8

Your Heart and Blood in Action.....................10

Open Wide! 12

Meet the Boss! 14

A Healthy Diet 16

Your Healthy Day 18

Rest, Repair, Renew... 20

Let's Investigate Drugs..................................... 22

Offspring 24

A Human Life Cycle 26

Extraordinary Bodies 28

Glossary... 30

Index... 32

The download button shows there are free worksheets or other resources available. Go to: www.rubytuesdaybooks.com/scienceKS2

Our Amazing Bodies

The human body is a finely tuned, natural machine that is capable of incredible actions and achievements.

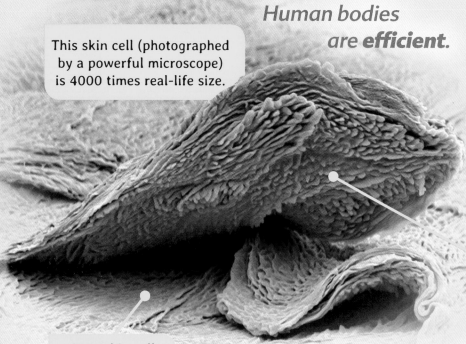

*Human bodies are **efficient**.*

This skin cell (photographed by a powerful microscope) is 4000 times real-life size.

An old skin cell about to fall off

New skin cell underneath

A Magnificent Machine

At this moment, your **heart** is automatically beating around 85 times each minute. Your **lungs** are taking in **oxygen** and your **stomach** is digesting your last meal – all without you thinking about it. Your eyes and brain are consciously reading this book and you will soon use **bones** and **muscles** to turn the page.

What We're Made Of

Your whole body is made of trillions of **cells** and they are growing and regenerating all the time. New bone cells grow as we grow bigger. New skin cells grow to repair a wound. In fact, the whole top layer of your skin regrows every month, shedding the old, dead cells at a rate of 40,000 every minute!

What Lives in Us?

Your mouth is home to billions of tiny friendly and unfriendly living things – **bacteria**. In fact, you have almost as many bacteria in your mouth as there are humans on Earth! Our whole bodies are home to these **microscopic** living things. Friendly bacteria in our stomachs help to digest our food. Unfriendly bacteria, or germs, can make us ill.

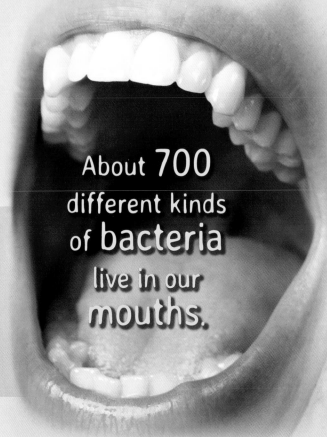

*Human bodies are **habitats**.*

About 700 different kinds of bacteria live in our mouths.

How Unique We Are

We are different from each other in many ways. But the one thing that makes each of us **unique** is our fingerprints. They form while we are in the **uterus** (womb) and stay with us for life, constantly regenerating our own special pattern of loops, whorls and arches. Even **identical twins** each have their own unique fingerprint.

Our fingerprints only change if they are cut or injured.

Loop

Whorl

Arch

*Human bodies are **individual** and **personalised**.*

How Tough We Are

Some humans test themselves to the limit in extreme sporting events. In the legendary Marathon Des Sables, competitors run a gruelling 250 kilometres in seven days – all in the Sahara Desert in scorching temperatures of more than 50°C!

The runners have limited food and water, which they must carry on their backs.

*Human bodies and minds are **strong, resilient** and **ambitious**.*

An Ironman Triathlon includes a non-stop 3.9-km swim, a 180-km cycle race and a 42-km marathon.

5

Structure and Strength

Inside your body is a strong framework of bones called a skeleton. Without it, you would just be a saggy bag of body parts that couldn't move!

Skeleton

Your skull protects your brain and gives your face shape.

Mandible

Joints For Movement

The places where your bones meet are called **joints**. Without joints, you would not be able to make movements. Knees and elbows are hinge joints that allow the bones to move in one direction and then go back again. Your bones are joined to each other by pieces of stretchy **tissue** called **ligaments**.

Cartilage

Bone

Ligament

Tough tissue called **cartilage** stops bones rubbing and wearing each other away.

Your ribs protect your heart, lungs and other vital organs.

There are 206 bones in the human body.

Spine

Humerus

Ulna

Radius

Hips and shoulders are ball and socket joints that make it possible to rotate your arm or leg.

Your thigh bones, or femurs, are the biggest bones in your body.

Patella

The **spine** is made of 33 bones called **vertebrae**.

Your spine, or backbone, runs from your head down to your bottom. It is highly flexible, allowing you to bend and twist. But it is also very strong, keeping your body upright and holding up your heavy head. Your spine protects your **spinal cord**, which is a large bundle of **nerves** that connect your brain to the rest of your body.

Tibia

Fibula

Between each vertebra is a disc of cartilage.

Your bones are moved, or pulled, by your muscles. Muscles are made of tough, stretchy tissue.

Muscles are attached to your bones by strong, stretchy cords called **tendons**.

Muscles

There are more than 650 muscles in your body. This diagram shows the names of some of the main muscles.

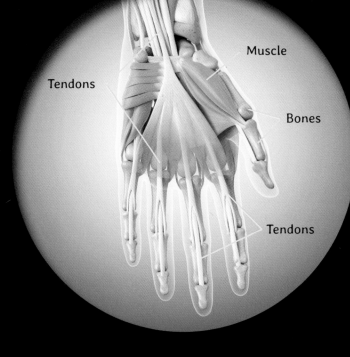

Tendons

Muscle

Tendons

Bones

Tendons

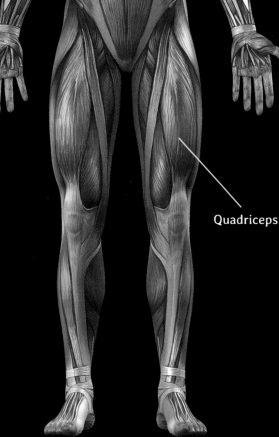

Trapezius

Pectoralis major

Deltoid

Biceps

Abdominal muscle

Quadriceps

Let's Test It

Lay your hand flat on a table with your palm down. Now wriggle your fingers. Can you see the cord-like tendons moving that attach the muscles to the finger bones?

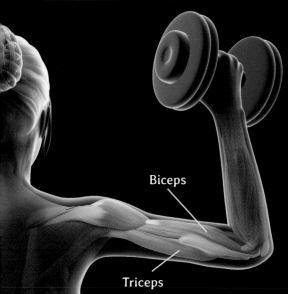

Biceps

Triceps

Contract and Relax

To make your bones move back and forth, muscles work in pairs. To make your arm bend, your biceps muscle contracts (or tightens) while your triceps muscle relaxes. To straighten your arm, your biceps relaxes and the triceps contracts and pulls on the bones.

The Digestive System

Everything you eat or drink goes on a long journey through your digestive system so that your body can take onboard every last bit of goodness.

Before you even take a bite, your mouth starts producing saliva, or spit.

Once your teeth get cutting and chewing, your tongue and teeth work together to mix the chewed-up food and saliva into a mushy ball called a bolus.

You swallow and the food goes down a long tube called the **oesophagus** into your stomach. Your stomach produces **digestive juices** that start to break down the food.

Muscles in your stomach churn and mix the juices and food until after about three hours, it looks like a thick milkshake.

From your stomach, the mixture moves on into a long, thin tube called the **small intestine**. More juices squirt from the walls of the intestine, breaking down the food so that its **nutrients** are released.

Oesophagus

Your stomach is like a stretchy bag or deflated balloon that increases in size as it fills with food.

The small intestine is a thin tube that's about 6 metres long.

The large intestine is a thick tube that's about 1.5 metres long.

Rectum

Anus

Along with nutrients, your small intestine soaks up the water that you drink and water from your food. Then the nutrients and water are released into your **blood**, ready to be carried wherever they are needed.

Now, all that's left of your food is solid waste that your body doesn't need. Muscles in the small intestine push the leftovers, or **faeces**, into a wider tube called the **large intestine**. Any remaining water is absorbed.

Then the faeces, or poo, move into the rectum and finally, leave your body when you go to the toilet through a hole called the anus.

An illustration of villi in the small intestine

Blood vessels inside the villi collect the water and nutrients.

It can take about 30 hours for food to move through your digestive system.

The inside walls of the small intestine are covered with millions of tiny parts called villi. The villi soak up nutrients and water and release them into your blood.

A mixture of bacteria (the colourful worm-like shapes) in faeces (greenish-brown substance).

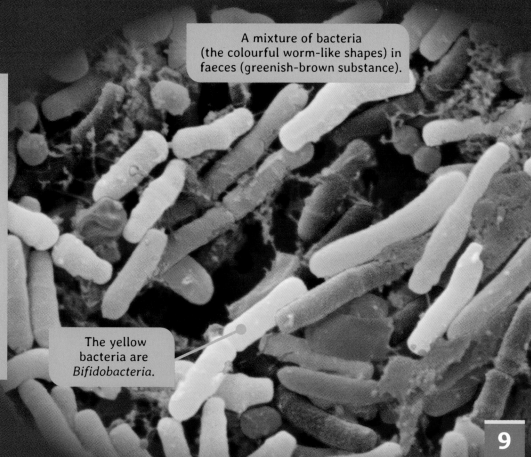

Friendly Bacteria

Your digestive system is home to trillions of bacteria that help digest food. *Lactobacteria* live in the small intestine. They help us break down foods such as milk and cheese. The large intestine is home to *Bifidobacteria*. They help us poo regularly and they kill off bad bacteria that can cause illnesses.

The yellow bacteria are *Bifidobacteria*.

Your Heart and Blood in Action

To keep working, the trillions of cells that make up your body need oxygen, water and nutrients. Together, your heart, blood vessels and blood are your circulatory system, which delivers all these things.

The Circulatory System in Action

The cycle begins when your heart pumps blood to your lungs, where it picks up oxygen.

The oxygenated blood travels back to the heart. Then the blood is pumped out to the rest of your body through a system of blood vessels called arteries. The blood delivers oxygen to any cell that needs it.

On its journey, the blood visits the digestive system to absorb water and nutrients from your food. These essential substances are delivered to your body, too.

As it flows through your arteries, the blood picks up unwanted **carbon dioxide** from your cells.

Once the deliveries and pick-ups are done, the blood travels back to the heart through a system of blood vessels called veins.

The heart pumps the blood back to the lungs. The blood drops off the carbon dioxide, picks up more oxygen and the cycle begins again.

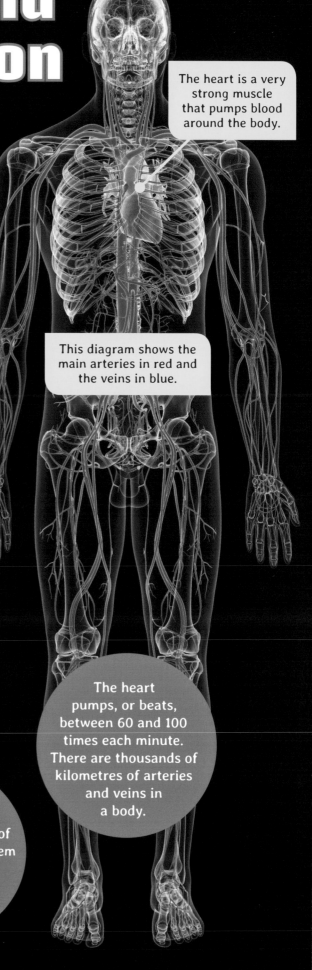

The heart is a very strong muscle that pumps blood around the body.

This diagram shows the main arteries in red and the veins in blue.

The heart pumps, or beats, between 60 and 100 times each minute. There are thousands of kilometres of arteries and veins in a body.

In real life, all blood vessels are red, the colour of blood. Looking at them through your skin makes them look blue.

In and Out

To create energy, your muscles and other body parts need oxygen. When you breathe in, air goes down into your lungs where the oxygen is absorbed into your blood. Unwanted carbon dioxide gas that has been collected by your blood leaves your body when you breathe out.

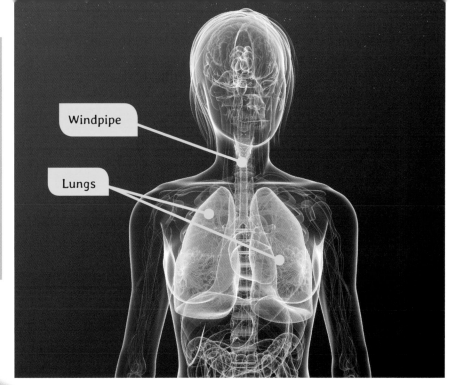

Windpipe

Lungs

Your Blood's Ingredients

The main ingredient in blood is a liquid called plasma, which carries water and nutrients to your cells.

Red blood cells deliver oxygen.

Bacteria

White blood cells attack harmful bacteria to fight off illness.

Platelets are cells that stick together like glue to form clots that stop blood flowing from cuts. As the clot dries, it becomes a crusty, protective scab to keep out bacteria.

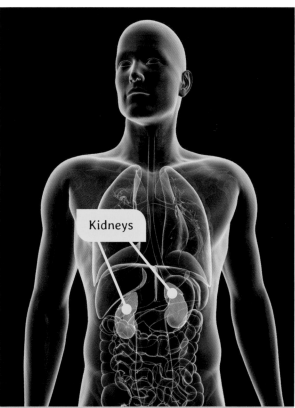

Kidneys

Your Kidneys at Work

As your body's cells use oxygen and nutrients, they create unwanted gases and chemicals. This waste is collected by your blood and carried to your kidneys. The kidneys remove the waste from the blood. They also turn any unwanted water into urine. When you go to the toilet, all the spare water and waste from the blood leaves your body as wee.

Four Types of Teeth

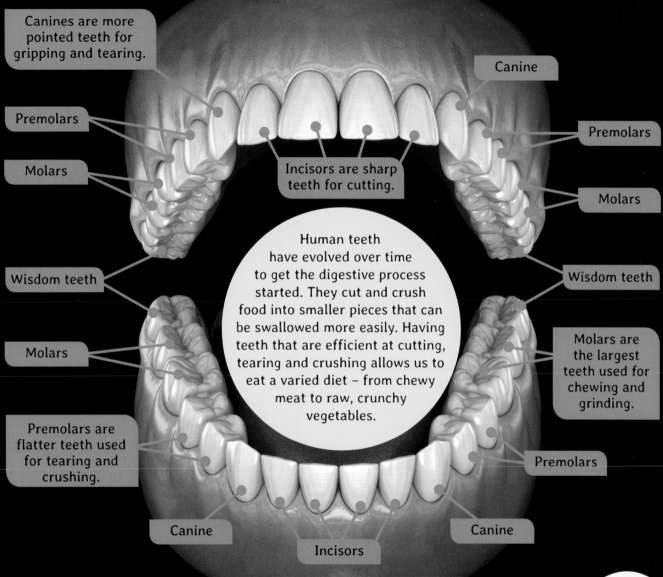

Canines are more pointed teeth for gripping and tearing.

Canine

Premolars

Premolars

Molars

Incisors are sharp teeth for cutting.

Molars

Human teeth have evolved over time to get the digestive process started. They cut and crush food into smaller pieces that can be swallowed more easily. Having teeth that are efficient at cutting, tearing and crushing allows us to eat a varied diet – from chewy meat to raw, crunchy vegetables.

Wisdom teeth

Wisdom teeth

Molars

Molars are the largest teeth used for chewing and grinding.

Premolars are flatter teeth used for tearing and crushing.

Premolars

Canine

Canine

Incisors

Most people grow four more adult teeth called wisdom teeth when they are about 20 years old. Dentists often have to remove these teeth as there may not be enough room for them.

Teeth are made up of dentine covered in shiny white enamel. Enamel is the hardest part of the human body. Unlike bones, teeth are not able to regrow cells and fix themselves if they are damaged or broken.

Inside a Tooth

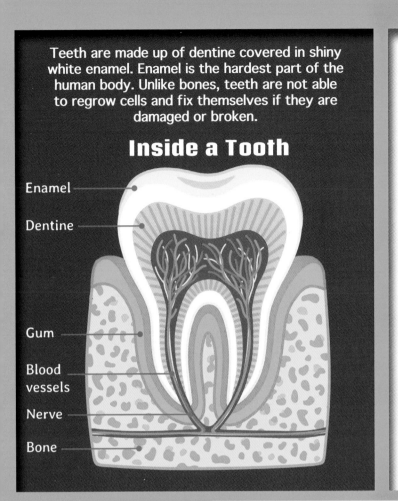

Enamel
Dentine
Gum
Blood vessels
Nerve
Bone

Top Tooth Tips

If you don't brush your teeth regularly, a sticky substance called plaque builds up on your teeth. Plaque contains millions of bacteria that will cause tooth decay and damage your gums. Eating too many sugary foods can cause tooth decay, too.

- Brush your teeth with toothpaste for two minutes, twice a day. One time should be before going to bed.

- Floss between your teeth to remove bits of food.

- Visit your dentist twice a year for a teeth and gum check-up.

Let's Investigate

Observe how cola can affect the shell of an egg and think about what this means for teeth that are exposed to lots of sugary drinks and no brushing.

Do sugary drinks damage our teeth?

Equipment:
- Hard-boiled egg (one with a light-coloured shell works best)
- Toothbrush and toothpaste
- Can of cola
- Empty jam jar
- Spoon
- Notebook and pen

Method:

1 Place the hard-boiled egg into the jam jar, cover with cola and leave for 24 hours.

What do you think will happen to the egg? Why? Write your prediction in your notebook.

2 Remove the egg from the cola with a spoon and carefully observe its shell.

What has changed? Record your findings.

3 Using the toothbrush and toothpaste, brush the egg as you would your teeth, rinsing afterwards. Record what the egg looks like after brushing.

What do your results show?

What do you think would happen to the egg if you left it for a week? A month?

How does this investigation support the idea that too many sugary drinks and not enough brushing are bad for our teeth?

Meet the Boss!

It may just look like a wrinkly, grey jelly, but your brain is in charge of everything that happens in your body.

Instructions from your brain travel down a long bundle of nerves in your spine called the spinal cord. Then they speed along nerves to every part of your body.

Your body parts send messages back to the brain along the nerves and spinal cord, too. For example: *Ouch! Remove the hand from that hot radiator!*

Brain

Spinal cord

This diagram shows the main nerves in the nervous system.

Nerves

Your brain has many different parts with different jobs.

When you read, turn the page of a book or talk, the cerebrum is at work. It also stores your memories.

The cerebellum controls balance and movement, and makes your muscles work together.

The brain stem controls all the things you don't have to think about like breathing and blinking. It keeps your heart pumping blood and tells your stomach to digest food.

Build a Body

Have a go at building your own life-size body using modelling materials. Think about which material will represent each part of your body the best.

Equipment:
- A large piece of paper as long and wide as you, or several pieces taped together
- A black marker pen
- Coloured pens, pencils or paints
- Scissors
- Glue
- Coloured paper, cardboard, tissue paper
- Modelling materials such as egg cartons, toilet paper tubes, felt, fabric, ribbon, straws, sponge, pipe cleaners, plasticine, balloons

Method:

1. Lie down on the paper and ask someone to draw an outline around your body in marker pen.

2. Using a pencil, map out where all of the main bones and muscles are.

3. Sketch out the main organs on pieces of paper, cut them out and place them in their correct positions.

4. Consider what materials might help show the function or texture of the organs, bones and muscles. Use your modelling materials to create these.

5. Label each body part, including information about its function.

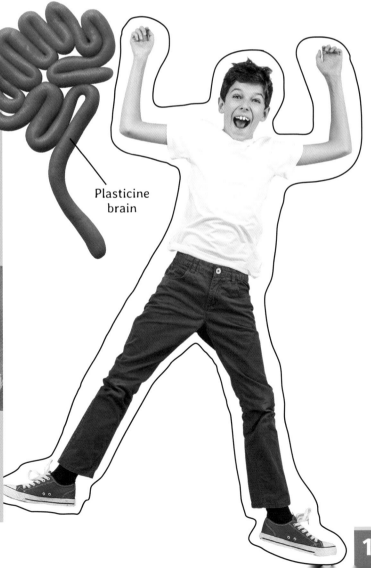

Plasticine brain

How to Make a Spine

The spine is made up of vertebrae, a spinal cord and cartilage. Together they are strong and flexible. To represent this on your model, you could use:

Spinal cord — pipe cleaner (long, strong and flexible)

Vertebrae — sections of egg cartons (strong and rigid)

Cartilage — sponge (soft and protective)

Make holes through the sponge and egg carton sections and thread them onto a long pipe cleaner (or several joined together).

A Healthy Diet

To have a healthy and balanced diet, we need to eat a range of foods from different food groups. This helps us to grow, gives us energy and makes us feel good.

The Main Food Groups

The foods we eat can be sorted into four main groups. Each group provides your body with the different nutrients it needs, including **vitamins**, **minerals** and **fibres**. This diagram shows how much of each food group should be on your plate at meal times.

We should eat foods from every group each day.

Starchy Carbohydrate Foods

- Main source of energy.
- Give us fibre which keeps our stomachs and intestines healthy by helping food move through our digestive system efficiently.

Pasta

Couscous

Potatoes

Bread

Wraps

Cereal

Protein Foods

- Help our bodies to grow and make repairs.
- Give us energy.
- Give us healthy skin.
- Keep our brains healthy.
- Vegetarians do not eat meat and vegans eat no animal products. Foods such as beans and nuts can help to build a balanced vegetarian or vegan diet.

Peanut butter

Meat

Eggs

Lentils

Tuna Fish

Beans

Dairy Foods and Substitutes

- Contain calcium for healthy teeth and bones.
- Give us protein and vitamins.
- Keep our brains healthy.
- Give us energy.
- Give us healthy skin.

Cheese

Milk

Fromage frais

Fats, Sugars, Salts

Sweets, chocolate, cakes, biscuits and crisps can be high in fat, sugar and salt. They can be part of a healthy diet as long as they are eaten in small quantities and infrequently. Eating too much of these foods can damage our teeth and make us overweight.

Some high-fat foods, such as unsalted nuts and olive oil, are highly nutritious and can be eaten in small quantities.

Biscuits

Nuts

Chocolate

Fruits and Vegetables

Peas

- Give us a range of vitamins.
- Help our bodies to fight off illnesses.
- Give us energy.
- Give us fibre which keeps our stomachs and intestines healthy by helping food move through our digestive system efficiently.

Cucumber

Vegetable kebab

Sweetcorn

Cherry tomatoes

Banana

Apple

Salad

We should aim to eat five portions of fruit and vegetables every day. This includes fresh, dried, frozen and canned fruit and vegetables. A drink of 150 ml of fruit or vegetable juice or smoothie also counts, but only once a day.

Strawberries

Broccoli

Grapes

Let's Experiment

Design a Super Salad!

Salads aren't limited to just lettuce, cucumber and tomato. Any mixture of raw and cooked vegetables can be the base of a salad. Look at the food groups diagram and make your salad healthy and balanced. You could add a starchy carbohydrate, such as cooked potatoes or pasta, and foods such as nuts, seeds, meat, cheese, eggs and even fruit.

Write a recipe for making your delicious super salad.

Your Healthy Day

All human cells contain water. In fact, our bodies are around 60 percent water! Your body needs water for making blood and for removing waste from your body.

Feeling **thirsty** is your body's way of telling you it **needs water!**

If you don't drink enough water you will get dehydrated. This can give you a headache or make you feel dizzy or tired. You may not be able to concentrate and might feel grumpy. Severe dehydration can make a person seriously ill and they may need hospital treatment.

Keeping Cool

Our bodies use water to stay cool. On a hot day or when we've been exercising, our bodies often sweat. As the mixture of water and chemicals dries on our bodies, it removes some of the heat.

Aim to drink **6 to 8 glasses** of water **every day.**

We can take in water from foods, too. Watermelons, cucumbers and tomatoes are around 90 percent water!

Milk and sugar-free fruit juice can also be part of your daily intake of liquids.

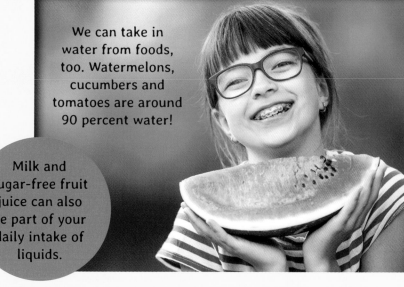

Let's Get Moving!

Eating well and drinking plenty of water are only part of a healthy lifestyle. It is vital that we also get moving and do exercise to stay physically and mentally healthy.

You should be physically active for at least one hour every day.

- Dance, exercise or martial arts class
- Swimming
- Brisk walk
- Play a sport
- Ride a bike

Equipment at your local park can be used for exercising.

Let's Design It

Can you create an intensive 15-minute exercise routine?

Plan it so you alternate between one minute of exercise and 30 seconds of rest. You could include star jumps, burpees, squats and jogging on the spot. Add some additional time for a warm-up before exercising and a cool-down afterwards. YouTube videos can be a good source of information and inspiration to get moving!

Your Healthy Day Checklist
It's fine to have treats and less active days now and then. But it's important to balance these times with days packed full of healthy choices. Are you on your way to ticking off these healthy targets for today?

3 balanced meals ✓
5 portions of fruit and veg ✓
6 to 8 glasses of water ✓
1 hour of physical activity ✓
2 lots of teeth brushing ✓
9 to 11 hours of sleep ✓

Rest, Repair, Renew

Humans spend about one-third of their lives asleep. As we sleep, our cells are repairing and renewing themselves and our body is busy getting ready for the next day.

When you are seven to 12 years old, you need nine to 11 hours of sleep each night. How much sleep do you get?

Nature Says It's Time for Bed

When it starts to get dark, your brain releases a chemical called melatonin into your body to make you feel sleepy. But if you spend time looking at a screen in bed, the bright light can make your body think it's daytime. This disrupts your sleep pattern and can make it hard for you to fall asleep.

- Try to go to bed at the same time each night so your body gets used to the routine.
- Make sure your bedroom is quiet and dark.
- Make your bed a place for sleeping – no TV or screens.

What Happens as I Sleep?

- Your heart slows down just a little and gets some rest, ready for lots of blood-pumping tomorrow.
- Your body uses its energy to make your bones and muscles grow.
- Your hardworking muscles rest.
- New cells grow and existing cells make repairs. Your skin grows faster to heal wounds.
- Your body produces chemicals to help fight off illnesses.

Let's Talk!

How might you feel at school if you don't have a good night's sleep? Will it affect your attention, learning or even your friendships? Why?

5 Stages of Sleep

Our bodies go through five stages as we sleep. These stages repeat in a 90-minute cycle throughout the night.

Stage 5
You are in REM (rapid eye movement) sleep. Your eyes may move about rapidly, making your eyelids flicker. You dream and your brain is nearly as active as it is during waking hours.

Stage 1
You're drifting off and if someone wakes you at this point, you may not even realise that you were asleep.

Stage 4
You are sleeping deeply and might be difficult to wake.

Stage 3
You are sleeping deeply and your body is relaxed and floppy.

Stage 2
Your breathing slows and your body temperature cools down.

Sleep helps your brain to concentrate, remember things, solve problems and help you feel happy.

Understanding Ourselves

Being mentally healthy means that we can feel and control a range of emotions, make and keep happy relationships and feel positive about ourselves. Sometimes, however, we feel less in control of our thoughts and emotions, which can make daily tasks tricky.

Think of a time when you found it challenging to control an emotion. Maybe you were feeling angry, sad or worried.

What were you *thinking* at the time? Was that true or helpful?

Where were you?
Who were you with?
Did this affect your emotions?

How did you *behave*?
Did this make it worse or better?

How did your body *feel*?
For example, did you feel tingly or tired?

If you were to relive that situation, what would you do or say differently to be more in control of your thoughts and emotions?

The word "drugs" can mean lots of different substances. Are all drugs bad for us? What are illegal drugs? Let's investigate!

Some drugs are used to make people better when they are ill. We usually call these drugs medicines. They can be bought from a pharmacy or prescribed by a doctor.

Some adults may choose to consume legal drugs such as caffeine, alcohol and nicotine, which is in cigarettes and e-cigarettes. In the UK, it is illegal to buy alcohol or cigarettes if you are under the age of 18.

Always take the right **dose** of any medicines you've been prescribed to make sure they do not cause further harm to your body.

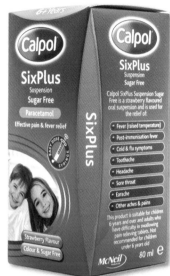

The correct dose of medicine to take is on the packet.

The Facts: Caffeine

Caffeine is a stimulant drug. This means it can make you feel more awake and alert. Coffee and tea naturally contain caffeine and it is added to energy drinks. However, drinking too much caffeine can alter your behaviour. You might feel panicky, anxious or grumpy. You may also find it difficult to sleep.

Energy drinks often contain lots of other harmful ingredients, including large amounts of sugar.

The Facts: Nicotine

Nicotine is a stimulant drug that is **addictive** and is found in cigarettes and most e-cigarettes. Many smokers say nicotine helps to stop them feeling stressed or anxious, but smoking poses severe health risks. When someone breathes in smoke from a cigarette, they breathe in more than 5000 chemicals. Many of these chemicals are poisonous and can cause cancer and other diseases.

5000 chemicals

Breathing in secondhand smoke from someone else's cigarette can mean you will also breathe in 5000 chemicals. In England and Wales it is illegal to smoke in a vehicle if someone under 18 is present.

The Facts: Alcohol

Alcohol is a depressant drug. This means it can slow down how you respond to things both physically and mentally. Some people say they feel happier and more relaxed when they drink alcohol. But it can also make you feel angry or sad, especially if too much is consumed. Some people can get addicted to alcohol. This means they drink too much and cannot stop. An alcohol addiction can make a person very ill, affect how they do their job and harm their relationships with family and friends.

I've Got a Hangover!

Drinking too much alcohol can leave you with a headache and sickness the next day. People call this a hangover.

The Facts: Illegal Drugs

ILLEGAL

Some people choose to take illegal drugs. Taking these substances is against the law because they can be highly dangerous. Some drugs can cause the user to forget who they are, see things that aren't there or make them extremely paranoid and anxious. Illegal drugs can sometimes cause death! These drugs are often highly addictive, causing some users to focus their whole lives on getting hold of drugs and taking them.

Offspring

The biological children of humans and the young of animals are known as offspring. Humans and other animal offspring have a mixture of DNA from their mother and father.

DNA carries all the information about how you will look, grow, function and reproduce. There is DNA in almost every cell in your body. DNA is long and thin and is coiled into structures called chromosomes within your body's cells.

Inside each of your cells that contain DNA there are 46 chromosomes — 23 from your mother and 23 from your father. This is why children look like their parents. Half of their DNA is from their mother and half from their father.

What does "biological offspring" mean? It means a child (or animal) that is related by birth to their mother and father. A child is not biologically related to a stepmum, stepdad or adopted parent, although they would still be part of their family.

A child might **inherit** skin colour, hair colour, height and many other characteristics from their parents.

DNA is short for deoxyribonucleic acid. A section of DNA looks like this.

Let's Talk!

Do you have biological siblings — brothers or sisters from the same parents? What features do they have that are similar to you? From which parent do you inherit those features? (If you don't have any biological siblings, observe a friend who does and try to answer the questions.)

All About Twins

Some mothers are pregnant with more than one child at a time. If they have two babies, the children are called twins. There are two different types of twins – fraternal twins and identical twins.

What Makes Fraternal Twins?

Sometimes a woman's **ovary** releases two **eggs** instead of one. Each egg is then **fertilised** by a different **sperm** from the man. The two eggs eventually become **embryos** that grow into babies. Each child has DNA from both parents just like any other biological siblings. However, the children are born at the same time and are the same age.

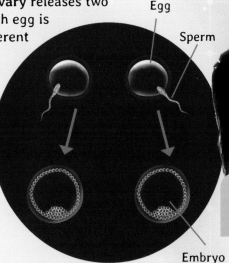

Egg

Sperm

Embryo

Fraternal twins can be a boy and a girl, two girls or two boys. They may look similar, as other brothers and sisters do.

What Makes Identical Twins?

Identical twins come from just one egg and one sperm. The fertilised egg quickly divides in two and grows into two children with exactly the same DNA.

Single egg and sperm

Egg divides

Embryo

Identical twins look the same and are always two girls or two boys.

A Human Life Cycle

Male Life Cycle →

Female Life Cycle →

Just like all **mammals**, humans begin their lives inside their mother's body.

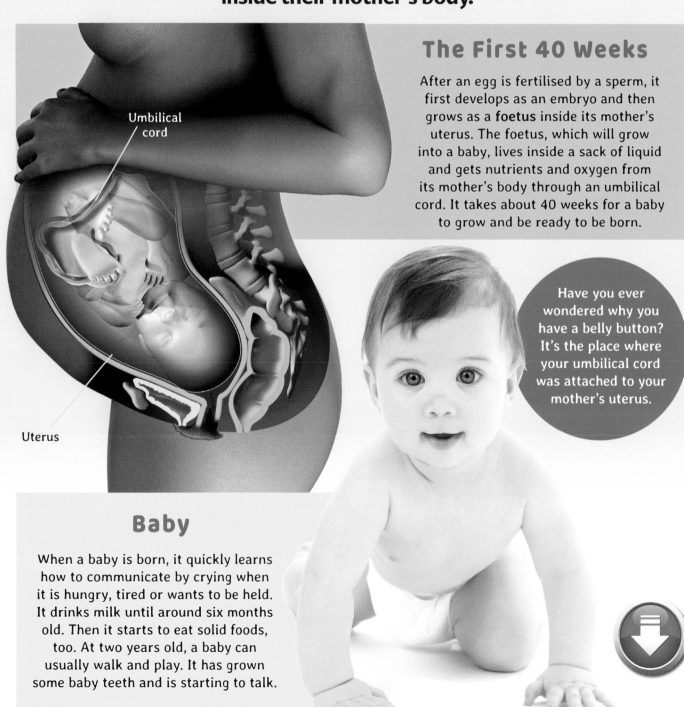

Umbilical cord

Uterus

The First 40 Weeks

After an egg is fertilised by a sperm, it first develops as an embryo and then grows as a **foetus** inside its mother's uterus. The foetus, which will grow into a baby, lives inside a sack of liquid and gets nutrients and oxygen from its mother's body through an umbilical cord. It takes about 40 weeks for a baby to grow and be ready to be born.

Have you ever wondered why you have a belly button? It's the place where your umbilical cord was attached to your mother's uterus.

Baby

When a baby is born, it quickly learns how to communicate by crying when it is hungry, tired or wants to be held. It drinks milk until around six months old. Then it starts to eat solid foods, too. At two years old, a baby can usually walk and play. It has grown some baby teeth and is starting to talk.

Childhood 3 to 12 Years

During our childhood years, we become more independent from our parents. We learn to go to the toilet by ourselves and learn how to get dressed. We go to nursery and then to school.

Adolescence or Teenage Years

Adolescence begins at about 12 years old or when our bodies start to show signs of changing into being an adult. At this stage, our bodies change in many ways as they get ready for adulthood and for reproduction. This time in our lives is called **puberty**.

Being an Adult

We become adults at about 18 to 20 years old. At this age we are usually at our peak fitness and our bodies are ready to have children if we want to. Having a healthy lifestyle can help us stay fit and strong as we grow older.

Being an Older Person

As we get older we may retire, which means no longer going to work. Some older people have trouble getting around, seeing, hearing or remembering things. But many older people take care of grandchildren, have lots of hobbies and are fit and active. As we get older our hair usually becomes thinner and turns grey or white.

Signs of Puberty

Boys

Develop body hair on their chest, armpits, face, legs and around their penis and testicles.

Voice deepens.

Shoulders get wider.

Muscles get bigger and stronger.

Boys and Girls

Skin gets oily and sometimes spotty.

Changes in emotions and mood.

Produce more sweat.

Grow taller.

Girls

Develop body hair on their armpits, legs and around their vulva.

Body becomes curvier and hips widen.

Breasts start to develop.

Monthly menstruation (period) starts.

Adult man

Child

Adult man

Adult woman

Teenager

Older people

Baby

Extraordinary Bodies

Living with a medical condition or disability can be challenging. But it doesn't stop people from achieving extraordinary things.

Jonnie Peacock

At the age of five, Jonnie Peacock contracted meningitis. This dangerous infection caused the tissues in his right leg to die. After having his leg amputated (removed), Jonnie had a prosthetic, or false, leg fitted to help him walk and run. Having always enjoyed running and competing at school, Jonnie was determined to be a successful Paralympian. He won gold medals at both the 2012 and 2016 Summer Paralympic Games in the 100-metre sprint.

Kelly Gallagher

Britain's first winter Paralympic Gold medallist, Kelly Gallagher, is a partially sighted ski racer. Reaching speeds of 120 km/h, Kelly fearlessly hits the slopes with a fully sighted guide to ensure she can safely navigate the course.

Jonnie Peacock in action in the Men's 100 metres at the London 2012 Paralympic Games

Kelly in action on the slopes

People are sometimes born with parts of their bodies working differently, or not as well as they might for other people. Sometimes a person gets ill or has an accident which means a part of their body no longer works as it did before.